D0004318

ONE SMALL STEP

Apollo 11 and the Legacy of the Space Age

Smithmark books are available for bulk purchase for sales promotion and premium use. For details, write or call the manager of special sales, SMITHMARK Publishers, 115 West 18th Street, New York, NY 10011.

Produced by SMITHMARK PUBLISHERS
115 West 18th Street, New York, NY 10011.

Design: Kay Schuckhart, Blond on Pond
Creative Direction: Kristen Schilo, Gato & Maui Productions

ISBN: 0-7651-1666-9

Library of Congress Catalog Card Number: 98-68329

Printed and bound in Hong Kong

10 9 8 7 6 5 4 3 2 1

To the three astronauts who dared to make a scientific dream come true,
we dedicate this book to Edwin "Buzz" Aldrin, Neil Armstrong, and Michael Collins.

Acknowledgments

A fond thank you to space lovers Steven Karchin and Gary Kraut at Alphaville, New York, for letting us borrow
some of their out-of-this-world toys, games, magazines, and movie posters, as well as for their keen cosmic knowledge.
Thanks also to Gwen Pitman from NASA's media services for providing us with NASA's archive photographs.
A special thanks to Frederick C. Durant III, the Conservator of the collection of Bonestell Space Art International, for giving
us permission to reproduce some of the magical moon images by space art pioneer, Chesley Bonestell.

Contents

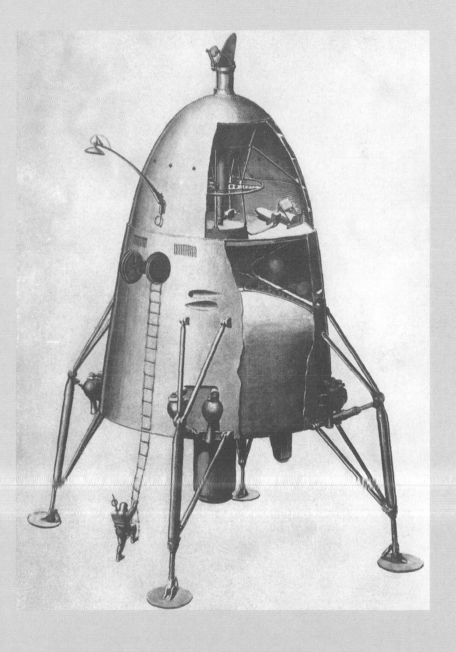

To the three astronauts who dared to make a scientific dream come true,
we dedicate this book to Edwin "Buzz" Aldrin, Neil Armstrong, and Michael Collins.

Acknowledgments

A fond thank you to space lovers Steven Karchin and Gary Kraut at Alphaville, New York, for letting us borrow
some of their out-of-this-world toys, games, magazines, and movie posters, as well as for their keen cosmic knowledge.
Thanks also to Gwen Pitman from NASA's media services for providing us with NASA's archive photographs.
A special thanks to Frederick C. Durant III, the Conservator of the collection of Bonestell Space Art International, for giving
us permission to reproduce some of the magical moon images by space art pioneer, Chesley Bonestell.

"That's one small step for man, one giant leap for mankind."

—Neil Armstrong, July 20th, 1969

Space Travel: From Fantasy to Reality

On the evening of July 20, 1969, people all over the world watched the grainy black-and-white televised image of Neil Armstrong as he took his first steps on the moon and uttered the words: "That's one small step for man. One giant leap for mankind."

"Giant leap" was perhaps both a boast and an understatement. Apollo 11's trip to the moon was more than the culmination of a decade-long, multi-billion dollar government mission. It reflected humankind's centuries-old desire to travel to the heavens, to go, in Jules Verne's words, "from the earth to the moon."

In the 1638 story "The Man In the Moon" by Domingo Gonsales, a flock of swans carries a man on a lunar mission. Cyrano de Bergerac described a trip to the moon that used solar "fuel" that was conveniently stored in bottles of morning dew. In 1827, American writer Joseph Atterly wrote of space travel via an antigravity substance called "lunarium." An antigravity device also powers the craft in H.G. Wells' 1901 tale, "First Men in the Moon."

BOTTOM: A scene from the 1902 film "A Trip to the Moon" directed by George Melies.

TOP: A 19th century print envisions a race of winged moon men escorting floating visitors.

OPPOSITE BOTTOM: Jules Verne even contemplated the effects of weightlessness on his travelers to the Moon.

OPPOSITE RIGHT: Less than a century later, Verne's fantastic vision became reality.

Of all the early moon travel fantasies, none was as influential as Jules Verne's 1865 novel "From the Earth to the Moon," and its 1870 sequel "Round the Moon." Rather than rely on pure fantasy, Verne calculated the power that an enormous cannon would require to propel a craft to the moon. Translated into many languages, Verne's novels inspired three early rocket pioneers: Russia's Konstantin Tsiolkovskii, Germany's Hermann Oberth, and the American Robert Goddard. All three men came to the conclusion that multistage rockets, not Verne's cannon, could reach the moon.

In 1919, the 37-year-old Goddard had the audacity to publish "A Method of Reaching Extreme Altitudes" about exploring the moon (and beyond) using rocketry. It earned him ridicule in the nation's press, including *The*

New York Times. The state of Massachusetts actually considered him a menace and forced him to take his experiments out of the state! Goddard wasn't deterred, and in 1926 he became the first man to perfect and launch a liquid-fueled rocket. So influential was Goddard's work on the space program that, in 1969, science fiction author Isaac Asimov proposed that the first words uttered on the moon should be, "Goddard, we are here!"

Yet it was in Germany that rocketry first found its most enthusiastic promoters. By 1932, members of the German Rocket Society had launched 87 experimental flights. Two of the society's early members, Willy Ley and Wernher Von Braun, went on to become early promoters of rocketry. After the Nazis came to power in 1933, Ley emigrated to America to found the American Rocket Society and popularize the notion that space travel was a real possibility, not a "Buck Rogers" fantasy.

Von Braun remained in Germany and was put to work by Hitler's military to develop the V-1 and V-2 rockets. When the first V-2 was launched on October 3, 1942,

THE CONQUEST OF SPACE

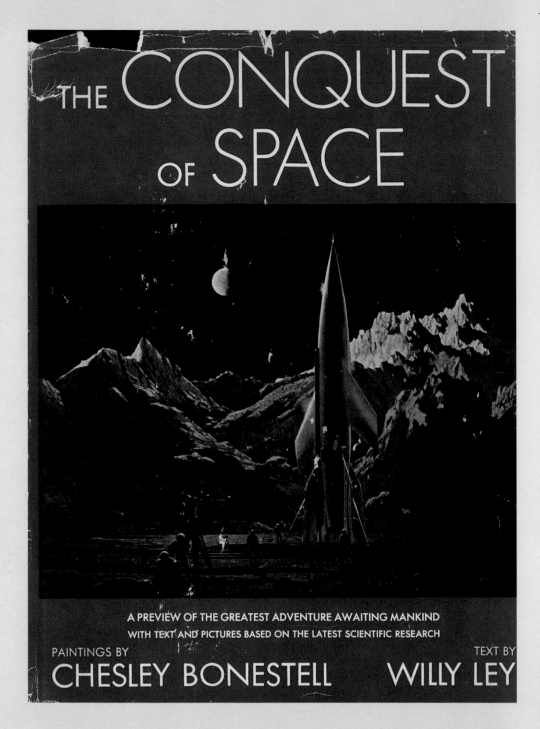

A PREVIEW OF THE GREATEST ADVENTURE AWAITING MANKIND
WITH TEXT AND PICTURES BASED ON THE LATEST SCIENTIFIC RESEARCH

PAINTINGS BY
CHESLEY BONESTELL

TEXT BY
WILLY LEY

**Rocket pioneer
Dr. Robert Goddard**

March 27, 1945, when the Peenemunde launch site was overrun by Allied forces. The rocket weighed 28,000 pounds, produced 56,000 pounds of thrust, and traveled faster than the speed of sound—it took a mere six minutes to reach London. At the end of the war, Von Braun and his fellow scientists were at work on a new generation of deadly rockets. Called the A-9 and A-10, they were the "America Rockets," to be launched from the west coast of France and aimed at New York.

As World War II ended, Von Braun and his team intentionally surrendered to the Americans instead of the Soviet forces so that they could help America develop better rockets, and hopefully begin a space travel program. But unfortunately for Von Braun, when he went to work at the U.S. Army's White Sands, New Mexico testing grounds, the two principal weapons of the coming Cold War, the atomic bomb and the intercontinental ballistic missile, were still in their infancy, and the United States government did not make rocketry a priority.

In contrast, the Soviet Union quickly improved upon its own W.W.II-era rocketry. By the end of the war the Soviets had a rocket with a range as far as the V-2—400 kilometers—but no enemy within that range. By 1949, under rocket scientist Sergei Korolev, the Russians had developed a T-1 missile capable of traveling 500 miles, and by 1955, the Soviet's R-7 rocket, with its 1.1 million pounds of thrust, had sufficient range to carry an atomic weapon that could attack the United States. It could also launch a satellite into orbit.

Dr. Wernher Von Braun

ABOVE: Willy Ley lent his name to a number of speculative books and essays about the real possibility of space flight.

RIGHT: "Zero Hour Minus Five" by Chesley Bonestell.

German military commander Walter Dornberger turned to Von Braun and declared, "Do you realize what we accomplished today? Today the space ship is born." Von Braun was briefly arrested by Hitler's SS when he was overheard saying that he was more interested in space travel than in weaponry.

The German rockets were indeed formidable. Hundreds of V-2 (or Vengeance Weapon 2) rockets hit London and other allied targets between September, 1944, and

TOP: "The Lunar Base" by Leslie Carr, based on a drawing by R. A. Smith.

ABOVE: Before Ley and Bonestell, space exploration was consigned to the realm of pulp fiction.

While Von Braun chafed under the American government's budgetary restraints and lack of vision, his old colleague Willy Ley joined the legion of scientists and science fiction writers who openly speculated about the possibility of space flight. Ley's 1947 book "Rockets: The Future of Travel Beyond the Stratosphere," and "Conquest of Space," his 1949 collaboration with illustrator Chesley Bonestell, popularized many space-related concepts, including earth orbit, the creation of a space station, and exploration of the moon. Royal Air Force veteran and science fiction writer Arthur C. Clarke published "Interplanetary Flight" in 1951 and later "The Exploration of Space" in which he argued that space travel was no longer a fantasy and "must be regarded as a matter beyond all serious doubt."

Others began to share Clarke's opinion and vision. In 1951, New York City's Hayden Planetarium sponsored a series of symposiums on space travel. There, Willy Ley discussed the reality of satellites, space stations, and lunar exploration. The next year, Ley's friend Von Braun used the symposium setting to propose a separate government agency to coordinate and promote space flight.

Editors from the popular weekly magazine *Collier's* attended the series and were inspired to run a series of articles on the exploration of space. Illustrated by Chesley Bonestell, these included essays by Von Braun and other experts and predicted a manned flight to the moon within 25 years. Soon other magazines, including *Life* and *Time,* provided Von Braun with a wider forum. Cartoonist and fantasy entrepreneur Walt Disney was so impressed with this new attention to outer space that he made space flight and exploration a key element of the "Tomorrowland" area of his Disneyland theme park which opened in 1954. Space exploration was also the focus of the Disney television program "Man In Space" which aired on March 9, 1955.

Hollywood Looks to the Stars

The efforts of Ley and *Collier's* to depict space travel as within the realm of possibility were not helped by Hollywood. Throughout the 1950s, the subject of space remained a B-movie fantasy. Among the few more serious films was "When Worlds Collide" (1951) in which scientists build a rocket ship ark to transport humanity to a safer planet when earth is faced with destruction by the passing planet Zyra. And the same year, the classic film "The Day the Earth Stood Still" depicted a thoughtful alien from space trying to warn humanity of the dangers of atomic warfare.

But most films stuck to fanciful science fiction. For example, acclaimed as "scientifically accurate" at the time of its release, the 1950 film "Destination Moon" featured a script co-written by science fiction writer Robert Heinlein—yet even this "serious" film saw fit to incorporate a Woody Woodpecker cartoon into its story. "Rocketship X-M" was United Artists' attempt to compete with "Destination Moon," and featured a plot about astronauts en route to the moon who are knocked off course by a meteor shower and land on a pink planet Mars which has been devastated by nuclear war. The 1951 film "Flight to Mars" used costumes left over from "Destination Moon" to depict a Martian civilization populated by beautiful women in short silver skirts. This theme was again picked up in the 1953 3-D film "Cat Women of the Moon," in which an expedition from Earth discovers a lunar race of tights-wearing telepathic beauties. It was remade (badly) in 1958 as "Missile to the Moon."

The ultimate '50s film about extraterrestrial beauty might be the 1958 hoot "Queen of Outer Space." It starred Zsa Zsa Gabor as a rebel girl from Venus who saves some traveling earth-men from her man-hating queen. Costumes for this low-budget movie were actually borrowed from the MGM hit "Forbidden Planet," which set Shakespeare's "The Tempest" on a faraway

Science fiction legend Robert Heinlein shared a co-writing credit on "Destination Moon." Bonestell was hired to make the film look "real."

AMAZING!
FORBIDDEN PLANET
IN CINEMASCOPE
STARRING
WALTER PIDGEON
ANNE FRANCIS
LESLIE NIELSEN
WITH
WARREN STEVENS
ROBBY, THE ROBOT
SCREEN PLAY BY CYRIL HUME
BASED ON A STORY BY IRVING BLOCK AND ALLEN ADLER
PHOTOGRAPHED IN EASTMAN COLOR
DIRECTED BY FRED McLEOD WILCOX
PRODUCED BY NICHOLAS NAYFACK
A METRO-GOLDWYN-MAYER PICTURE

SNEAK ATTACK! Earth Battles Outlaw Planet!

BATTLE IN OUTER SPACE

IN EASTMAN COLOR

with
RYO IKEBE · KYOKO ANZAI · LEONARD STANFORD · HAROLD CONWAY · GEORGE WHYMAN · ELISE RICHTER
Screenplay by ANINICHI SEKIZAWA Based on a story by JOTARO OKAMI Directed by INOSHIRO HONDA Produced by TOMOYUKI TANAKA
Special Effects by EIJI TSUBURAYA Filmed in TOHOSCOPE · A TOHO PRODUCTION A COLUMBIA PICTURES RELEASE

'right ©1960 Columbia Pictures Corp. Country of Origin U.S.A.   1   Property of National Screen Service Corp. Licensed for display only in connection with the exhibition of this picture at your theatre. Must be returned immediately thereafter.   60/

world that is visited by earthlings in a flying saucer. For all of its special effects and lavish production values, it was closer to the science of Elizabethan England than the reality of 1956.

Even as Russia's Sputnik made space flight a reality in 1957, most space films remained fantastic, crude, and exploitative. Released only two years before Yuri Gagarin's historic flight (see page 19), "First Man Into Space" (1959) depicted a returning astronaut who is turned into a blood-thirsty creature by space dust. The film "12 to the Moon" (1960) was no better. In it, moon-men threaten to freeze earth with a ray gun. But what could be expected from the director of "They Saved Hitler's Brain" and the "Phantom Planet?"

Perhaps the most unusual space film came from outside the Hollywood orbit entirely. "First Spaceship on Venus" (1960) was a Polish-East-German co-production about an international space probe of Venus. It was also probably the first film to show a black astronaut.

Fear of saucers, fear of monsters, fear of women: space movies from the '50s reflected the overriding fear of atomic war that dominated politics and pop culture during that decade. The 1960 movie "Rocket Attack U.S.A." (from the aptly named Exploit Films) faced those fears head-on and without subtlety. After a U.S. agent tries to steal Sputnik secrets, the Soviets respond with a missile attack that completely destroys New York City. The film concludes with the corny (and yet haunting) device: "Don't let this be—The End!"

OPPOSITE: "Forbidden Planet." MGM's remake of Shakespeare's "The Tempest" in outer space.

LEFT: Cold War fears were never far from the surface of 1950s and 1960s sci-fi dramas.

BOTTOM: The astronaut in "First Man into Space" returned to earth as a monster.

IT LEAPS AHEAD OF THE HEADLINES!

M-G-M presents

FIRST MAN INTO SPACE

Starring MARSHALL THOMPSON and MARLA LANDI
Screenplay by JOHN C. COOPER and LANCE Z. HARGREAVES
Produced by JOHN CROYDON and CHARLES F. VETTER, JR. · Directed by ROBERT DAY

The Shock of Sputnik

Soviet space efforts were not limited to the realm of film entertainment. Both the U.S. and the U.S.S.R. announced their intentions to launch an earth-orbiting satellite during the International Geophysical Year of 1957. As early as 1954, Wernher Von Braun had argued that the U.S. had the means to launch a satellite using the Redstone Rocket, an enhanced version of the V-2. "It would be a blow to U.S. prestige if we did not do it first," Von Braun argued. Once again, he proved a visionary. The U.S. government not only ignored Von Braun, but decided to entrust its satellite efforts to the Navy's less dependable Viking Missile project. In 1956 when Von Braun's Army rocket team launched his four-stage Jupiter rocket, he was strictly forbidden to include a satellite. Instead, the Army filled the fourth stage with sand.

And then, on October 4, 1957, the Soviet government announced that they had launched their first satellite, named Sputnik, Russian for "fellow traveler." Americans were in shock. They had always felt protected by two vast oceans, and now a Soviet projectile was traveling over American air space every hour. Many saw Sputnik as no less a threat than a second Pearl Harbor.

Secure that America had overwhelming nuclear superiority, President Eisenhower tried to downplay the significance of Sputnik or the need to match or equal the Russians. An advisor commented that Ike had no interest "in an outer space basketball game." But few felt reassured. In fact, Eisenhower's attitude convinced them that the old General was out of touch with a new and frightening world. The Soviets launched Sputnik II on November 3, 1957. This time it carried a dog, named Laika. But America's humiliation was not yet complete. In December 1957, the Navy's Vanguard rocket, carrying a four-pound satellite (Sputnik weighted more than 200 lbs.) exploded in front of the assembled media. The headline of the *London Daily Herald* read "Oh, What a Flopnik."

Finally, on January 29, 1958, Von Braun's team put America's first satellite, Explorer I, into outer space using a Jupiter C rocket. But the Sputnik embarrassment had a positive side, as it had galvanized the U.S. government into action. On October 1, 1958, the National Aeronautics and Space Administration, or NASA, was created to coordinate American space efforts. Space would no longer be reduced to a rivalry between the Army and the Navy; rather, it was now a major American priority.

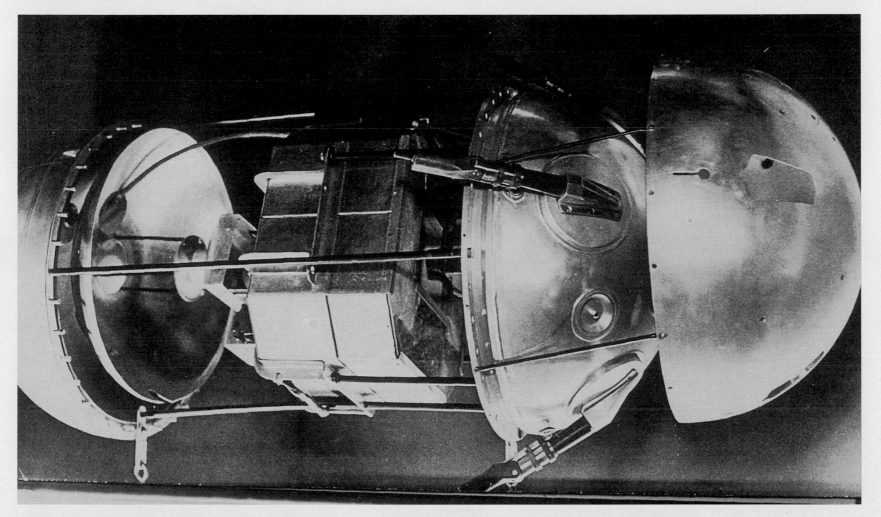

By September 1958, the Soviets had launched the first of several probes towards the moon itself. While they failed to successfully reach their target, these missions indicated that the Russians had rockets capable of reaching the 25,000 mph speed necessary to leave the earth's gravity. By the end of 1959, Soviet Luna satellites were sending back pictures of the dark side of the moon.

NASA completely captured the American imagination on April 9, 1959 with the introduction of the Mercury 7 astronauts, the first men scheduled to enter outer space. All seven had extensive military experience, including aerial combat during World War II and the Korean War. They were treated like conquering heroes by an adoring press. *Life* magazine was singularly enthusiastic in its celebration of their vigor and bravery. With recent images of exploding American rockets still fresh in their minds, the public was well aware that these gung-ho, crew-cut veterans were putting their lives at risk.

Despite all the hype surrounding the Mercury astronauts, the Eisenhower Administration seemed less than enthusiastic about making a commitment to space. Even after John F. Kennedy was elected president in 1960, Eisenhower publicly questioned the use of large funds for space exploration, calling it "a mad effort to win a stunt race."

However, having won the presidency with the promise to "get America moving again," Kennedy quickly dubbed his Administration "the New Frontier," a clear allusion to the frontier of the heavens. But Kennedy had barely been in the White House three months when the Soviets launched the Russian "cosmonaut" Yuri Gagarin into orbit on April 12, 1961, making him the first man to orbit the earth. Once again, America had been beaten, and humiliated, in space. When NASA spokesperson Lt. Col. John "Shorty" Powers was awakened by a reporter at 3 a.m. for a comment, his reply was terse, and telling. "If you want anything from us, you jerk, the answer is we are all asleep!"

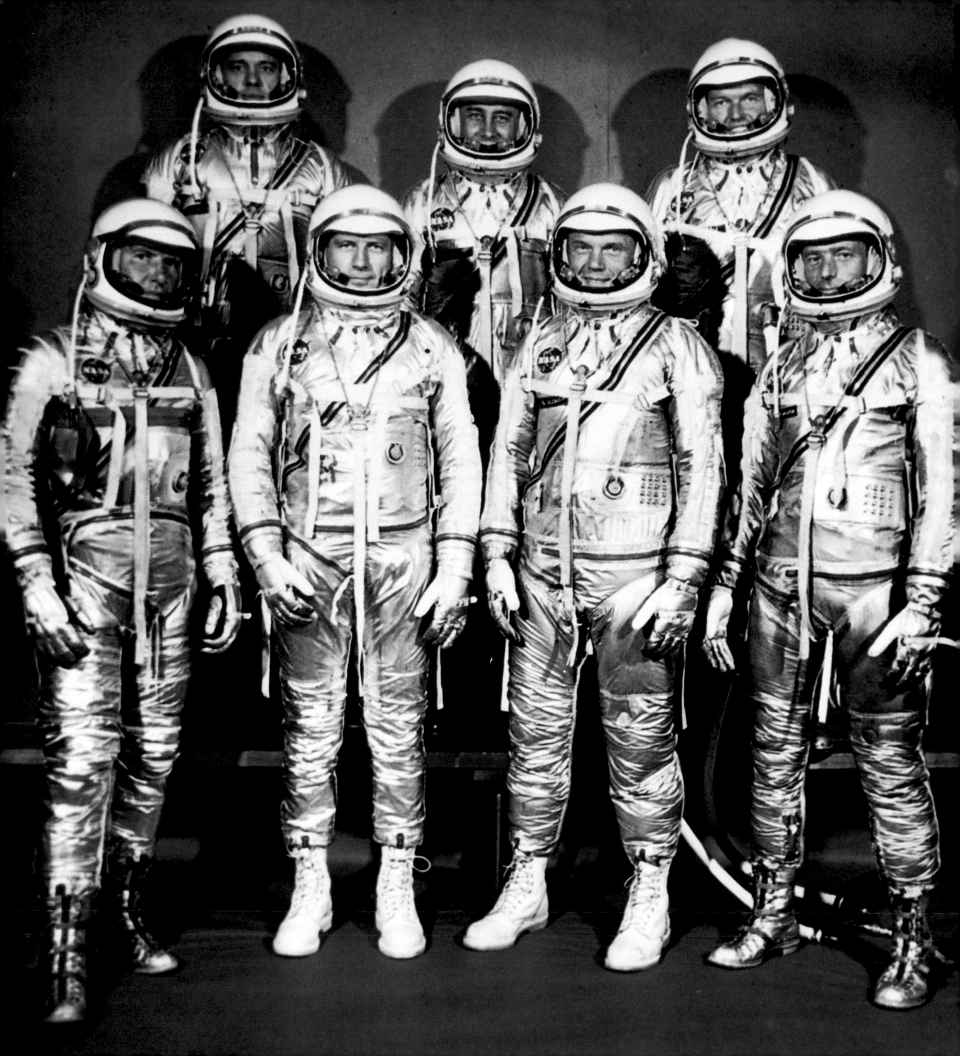

The Kennedy Challenge

American astronaut Alan Shepard's flight into space on May 5, 1961 proved to be a morale booster, as a Redstone rocket shot his Freedom 7 capsule on a 304 -mile sub-orbital flight. *Time*'s coverage was filled with ominous references to the Cold War and the real fear of World War III. "With Shepard rode the hopes of the U.S. and the whole free world in a period of darkness." But metaphors aside, it was hard to hide the fact that Shepard's flight was really too little and too late, as it lasted a mere 15-and-a-half minutes. In contrast, Gagarin's Vostok capsule had circled the globe. There was no comparison.

Against this dire background, Kennedy's bold May 25, 1961 declaration to "put a man on the moon and return him safely to earth before this decade was out," was a political and strategic masterstroke. With a few paragraphs, Kennedy changed the entire equation in space, and gave a boost to America's pride during a low period of the Cold War. America was no longer playing catch-up, but dedicated to a mission to reach the moon on a fixed timetable.

Kennedy committed the nation to the moon project during a special message to Congress on May 25, 1961. His remarks make clear that the lunar mission was very much a part of the Cold War fight for the hearts and minds of the world's population.

"If we are to win the battle that is going on around the world between freedom and tyranny, if we are to win the battle for men's minds, the dramatic achievements in space which occurred in recent weeks should have made clear to us all, as did the Sputnik in 1957, the impact of this adventure on the minds of men everywhere who are attempting to make a determination of which road they should take. . . . We go into space because whatever mankind must undertake, free men must fully share."

Then he added: "I believe this nation should commit itself to achieving the goal, before this decade is out, of landing a man on the moon and returning him safely to earth. No single space project in this period will be more impressive to mankind, or more important for the long-range exploration of space; and none will be so difficult or expensive to accomplish."

America's president had committed NASA and the nation to a moon mission. Now they just had to figure out how they would do it. There were three competing theories on how best to get to the moon and back:

Direct Ascent. While many sci-fi books and films envisioned a single rocket flying to the moon, landing, and then flying back, this scenario was simply out of the question. The giant NOVA rocket required for this mission was years away from flying.

Earth Orbit Rendezvous. Some argued that all of the spacecraft components necessary for a moon mission should first be sent into earth orbit, and then assembled in space for the long flight to the moon. While initially popular, this would require several rocket launches.

Lunar Orbit Rendezvous. This called for a the spacecraft to fly directly to the moon. Once in lunar orbit, a smaller landing craft would detach and descend to the lunar surface. After considerable debate, this third strategy finally carried the day, and was announced on November 7, 1962.

OPPOSITE: The original Mercury Project astronauts. Front row from left: Walter M. Schirra, Jr., Donald K. Slayton, John H. Glenn, Jr., and Scott Carpenter. Back row from left: Alan B. Shepard, Virgil I. "Gus" Grissom and L. Gordon Cooper.

BELOW: President John F. Kennedy.

"If we are to win the battle that is going on

around the world between freedom and tyranny, if we are to

win the battle for men's minds, the dramatic achievements in

space which occurred in recent weeks should have made

clear to us all, as did the Sputnik in 1957, the impact of this

adventure on the minds of men everywhere who are attempting

to make a determination of which road they should take . . . .

We go into space because whatever mankind must undertake,

free men must fully share."

John F. Kennedy, May 25, 1961

Mercury and Gemini

At the time of Kennedy's commitment to the moon landing, America's manned space experience consisted of Astronaut Shepard's 15-minute sub-orbital ride aboard Freedom 7. Gus Grissom's flight in July, 1961 lasted a few seconds longer. On February 20, 1962, astronaut John Glenn became the first American in orbit. His three spins around the globe in Friendship 7 were marred by serious concerns that the heat shield had detached. Without the shield, his craft would incinerate upon reentry into earth's atmosphere. But these fears were unfounded, and Glenn splashed down to a hero's welcome. In fact, if there were any doubts that the American public would embrace the space program they were dispelled by the tumultuous reaction to John Glenn's flight, which included a ticker tape parade up New York City's famed Broadway, and an address to a Joint Session of Congress.

Three more Mercury astronauts—Scott Carpenter, Walter Schirra, and Gordon Cooper—would orbit the globe three, six, and 22 times in 1962 and 1963. Project Mercury had proven that NASA could send men into space. But there was still a long way to go to the moon. Several preliminary space tasks had to be accomplished before a lunar landing. The first task was for one spacecraft to locate another spacecraft in orbit, maneuver towards that craft, and then successfully rendezvous and physically connect or "dock" with the second craft. The second task was for an astronaut to leave his cockpit and work outside the spacecraft. The third task was to test the reaction of the human body to an extended stay in outer space.

Nine Gemini missions between March 1965 and November 1966 achieved these goals with extraordinary efficiency. During the Gemini 4 flight in June 1966, Ed White became the first man to perform an extra-vehicular activity (EVA) or "walk in space," as it was called at the time. The next mission, Gemini 5, established a new

world's record for longevity, orbiting a total of 120 times in eight days.

By mid-1965, NASA was no longer playing catch-up with the Soviets. Launched within days of each other, Gemini 6 and 7 successfully linked up in orbit for the first space rendezvous. The mission of Gemini 7 lasted almost two weeks.

Captained by Astronaut Neil Armstrong, the Gemini 8 mission successfully docked with the unmanned Agena spacecraft. Four additional missions experimented successfully with docking procedures and EVAs. During Gemini 12's mission, astronaut Edwin "Buzz" Aldrin spent more than five hours working outside his orbiting craft.

In addition to the manned Gemini missions, NASA launched a series of lunar orbital satellites to map the lunar surface, measure radiation, and study the meteor bombardment of the lunar surface. Although rarely as celebrated as "flesh-and-blood" space missions, these unmanned probes proved essential to the Apollo Program and the eventual manned exploration of the moon.

ABOVE: President Kennedy hails astronaut John Glenn at a ceremony held at Cape Canaveral in March, 1962.

Pop Goes the Space Race

RIGHT: Mitch Miller provided the orchestration for the 1951 theme music to "Tom Corbett Space Cadet."

"Tom Corbett Space Cadet" puzzle circa 1950.

BELOW: Metal toy rockets made in Japan fueled many a child's space fantasies.

By the mid-1960s, the Mercury and Gemini space missions had made space travel a predictable part of American television news and entertainment. While grade school students in the 1950s were trained to "duck and cover" in case of atomic attack, their 1960s counterparts were taken to the school auditorium to watch televised space launches and splash downs.

With real space heroes being created on a regular basis, TV shows and movies had to develop a more realistic picture of astronauts. Gone were the fantastic plots and low-budget wonders of such early '50s shows as "Captain Video and His Video Rangers," "Space Patrol," or "Tom Corbett—Space Cadet." By the mid-'60s, space travelers were depicted as just regular guys.

"I Dream of Jeannie" (1965-70) may have had a title character right out of the "Arabian Nights," but her astronaut "master" Captain Nelson and his sidekick Major Healey were so everyday they were almost banal. Another "wacky" space comedy, "It's About Time," had two astronauts landing among cavemen when their craft took a wrong turn.

And clearly inspired by the style of the U. S. space program, the Irwin Allen-produced "Time Tunnel" (1966-67) depicted a secret NASA-like project to send men backward and forward through history.

The two best remembered space programs from the 1960s were far more fantastic than factual. "Lost in Space" borrowed plot lines from *The Swiss Family Robinson* and a character, Robby the Robot, from the 1956 film "Forbidden

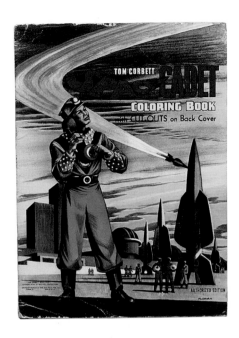

Planet," to depict a clean-cut American family marooned in a strange space world (along with that sneaky Dr. Smith). It was set in the far-off year of 1997.

Set in the 23rd century, "Star Trek" was more of an anthology for speculative science-fiction writing than a realistic view of space travel circa 1966. Like the 1968 movie "Planet of the Apes" and its sequels, "Star Trek" used space travel as a backdrop for commentary on '60s-era society. Its low-budget sets, numerous monsters, and scantily clad women were a visual throwback to the exploitation films of the '50s and before. At the same time, Captain Kirk's gung-ho nature and the show's motto, "to boldly go where no man has gone before," was very much

in the spirit of Kennedy's New Frontier. Despite its emergence as a cult classic, "Star Trek" never gained a large audience. It was canceled on September 2, 1969, only weeks after Armstrong's walk on the moon.

Director Stanley Kubrick's 1968 film "2001: A Space Odyssey" emerged as the most serious and influential space film of the decade, and perhaps ever. Based on the novel by the space visionary Arthur C. Clarke, the movie used dazzling special effects and balletic camera work to depict a very realistic vision of everyday life in space, including a space station, interplanetary probes, and regular flights to a permanent base on the moon.

TOP LEFT: "Tom Corbett" coloring book from 1950.

ABOVE: The TV show "Lost in Space" was set in the far-off year of 1997.

LEFT: A 1952 lunch box.

AN UNUSUAL AND IMPORTANT MOTION PICTURE FROM THE PEN OF PIERRE BOULLE, AUTHOR OF "THE BRIDGE ON THE RIVER KWAI"

20th CENTURY-FOX PRESENTS

CHARLTON HESTON

in

AN ARTHUR P. JACOBS PRODUCTION

PLANET OF THE APES

Complete with perky stewardesses, bad in-flight meals, space toilets, and space phones, the film employed real brand names and situations to give it an aura of nonfiction. It's no exaggeration to think that, based on the progress made in the eleven years between Sputnik and Apollo, a movie-goer in 1968 might expect that the situations depicted in "2001" could actually be realized in the next 30 years, if not sooner. Ironically, while we are nowhere near to colonizing the moon or travel-ing to Jupiter, three of the major institutions referred to in the film—Pan Am airlines, Bell Telephone, and the Soviet Union—have ceased to exist. And while there is no Clavius base on the moon, it *is* the name of the production company owned by actor, director, and space buff Tom Hanks.

For all of its realism, "2001" was promoted (and is still remembered) as a mind-altering experience. Movie posters touted it as "the ultimate trip," an obvious reference to the psychedelic pop culture of the period. Its story of an enigmatic monolith and a paranoid computer would provide grist for heady conversations for decades to come. The film reflected a radical change in the perception of space adventure. No longer was it a scientific journey or conquest, but a metaphor for a trip into the human mind. During the 1960s, a whole new generation was coming of age that seem far more interested in inner space than outer space.

OPPOSITE: Like The TV show "Star Trek," the film "Planet of the Apes" used an outer space setting to comment on contemporary social issues.

OPPOSITE BOTTOM: A space shoot-em-up game from 1958.

LEFT: The Steve Scott board game from 1952.

BOTTOM: Stanley Kubrick's "2001: A Space Odyssey" was sold as "the ultimate trip."

The Apollo Program Takes Shape

While Gemini made headlines, Wernher Von Braun and his rocket team continued to work on the huge Saturn 5 booster rocket. An earlier version, the Saturn 1, had made ten experimental flights between 1961 and 1967, launching unmanned Apollo modules and three enormous Pegasus satellites.

For all of its size, power, and complexity, the Saturn 5 was essentially just a very big rocket, a giant version of Von Braun's V-2 or Goddard's early prototype. Building the huge Saturn was a process of engineering, not invention.

On the other hand, the lunar excursion module (LEM), later shortened to lunar module (LM), was another matter. Nothing like it had ever been built before, or since. Once NASA decided on the lunar orbital mode of exploration, it needed a craft that could land softly on the lunar surface, provide housing for the astronauts while on the moon, and enable them to take off from their lunar base and rendezvous with their command module.

The LM consisted of two main parts, the ascent stage and the descent stage. The descent stage housed the rockets and fuel required to take the craft out of lunar orbit and safely land on the moon. It featured four spindly "legs," each with a 37-inch foot pad with sensitive probes that signaled the ship's engine to shut off when they hit the lunar surface. The legs gave the LM the appearance of an ungainly insect. Small wonder the Apollo 9 team, the first to fly the LM in earth orbit, christened their LM "Spider."

The ascent stage was the astronauts' lunar home. Just over 12 feet high and 14 feet wide, it was large enough for two astronauts to stand in as they guided the craft to a smooth landing. Two-foot-square triangular windows were positioned so the astronauts could see where they were going. This visibility proved crucial during the touch-and-go descent of the Eagle on the Apollo 11 mission.

The descent stage had a powerful rocket with an adjustable throttle which allowed astronauts to slow the LM down from orbital speed to a near dead stop. If need be, the LM could hover over the surface like a helicopter. In fact, the Eagle's landing was so soft its legs failed to depress as much as expected, making Armstrong's first descent longer and more awkward than planned.

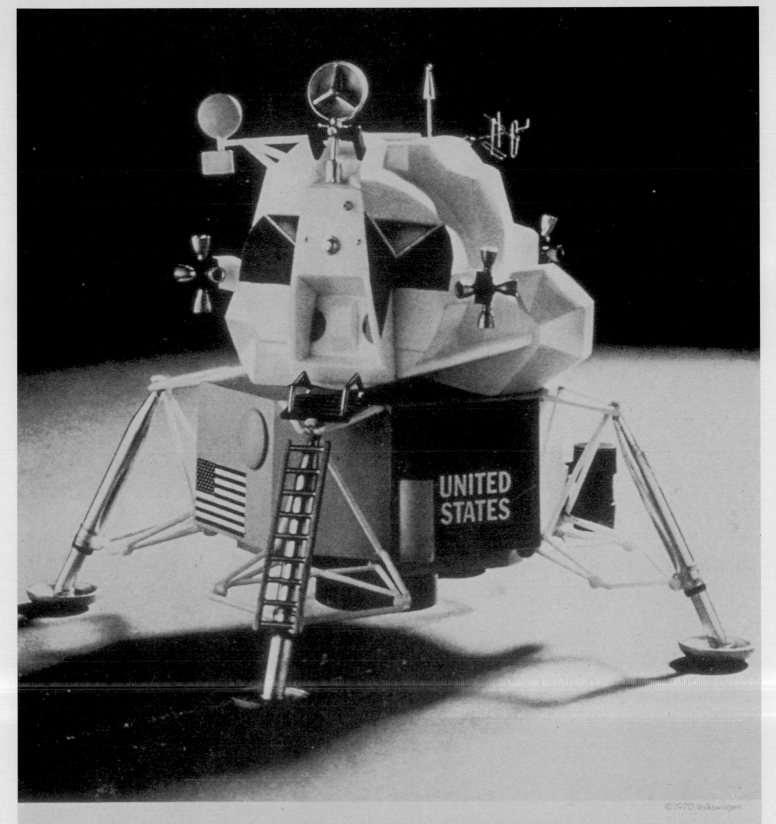

It's ugly, but it gets you there.

Upon liftoff, the descent stage became a mere launching pad for the ascent stage. Its much smaller engine had no throttle, but provided a steady 3,500 pounds of thrust capable of returning the LM to lunar orbit and rendezvous with the command module.

Engineers at the Grumman Aerospace Company on Long Island, New York realized that the moon's weaker gravity would enable them to make a lighter craft with lighter materials. Much of the LM was made by hand, with technicians taking a block of metal and milling it until it fit the blueprint. One engineer was brutally frank about its lack of beauty, claiming it had "a shape that at first looked ridiculous and looked more and more ridiculous as we worked on it." In fact, Volkswagen used a full-page photograph of the LM in some of their advertisements for their popular but homely compact car, with the fitting slogan, "It's ugly, but it gets you there."

Meanwhile, the development of the teardrop-shaped Apollo spacecraft was plagued with complications and delays. At one point, astronaut Gus Grissom was so upset with the number of modifications and engineering changes taking place that he hung a lemon outside his flight simulator to protest his working conditions. And then, on January 27, 1967, tragedy stuck. During a simulated countdown of Apollo 1, a spark ignited the pure oxygen atmosphere of the spacecraft, causing it to burst into flames and instantly kill the crew of Gus Grissom, Ed White and Roger Chafee. Project Apollo would remain grounded for the next 19 months.

On November 9, 1967, the huge 363-foot-tall, three-stage Saturn 5 rocket was successfully launched, carrying the unmanned Apollo

4 capsule. The rocket re-quired to send Apollo to the moon was finally ready. Two more unmanned Apollo flights would ensue before the Apollo 7 was launched by a Saturn 1B rocket to test the command module in earth orbit. Captain Walter Schirra pronounced it a "first class space mobile."

With only one manned Apollo flight behind it, NASA took a remarkable gamble with Apollo 8. While it was to be the first mission launched from the huge Saturn 5, it was originally intended to remain in earth orbit. Instead, NASA sent astronauts James A. Lovell, Frank Borman, and William A. Anders on a trip around the moon.

The spacecraft entered lunar orbit on Christmas Eve and spent 16 hours circling the moon. The three astronauts on board became the first humans to feel the moon's gravity and to see the earth in its entirety from a point faraway in space, as well as the first to behold the hidden "dark side" of the moon and to witness an "earth rise." The televised image of the blue planet set against the cold vacuum of space also had a profound effect on the millions of viewers back on earth. As the space crew's TV cameras panned over the desolate lunar surface, the astronauts read verses from the first ten chapters of the Book of Genesis, concluding with: "And God called the dry land Earth, and the gathering together of the waters called he Seas. And God saw that it was good. And from the crew of Apollo 8, we close with, good night, good luck, a Merry Christmas, and God bless all of you, all of you on the good earth."

After the drama of Apollo 8, the next two missions were comparatively workmanlike. Apollo 9 was the first manned flight to take the lunar excursion module along. It also instituted the Apollo program tradition of having the crew select names for the command and lunar modules. Astronauts James McDivitt, David Scott, and Russell Schweickart called their command module "Gumdrop" and the LM "Spider." Confined to earth orbit, they separated Spider from Gumdrop, jettisoned their descent stage, and docked again after six hours.

Apollo 10 astronauts Thomas Stafford, John Young, and Eugene Cernan took their cue from the comics pages and

named their command module "Charlie Brown" and their lunar module "Snoopy." Charlie Brown and Snoopy entered lunar orbit on May 22, 1969. During their separation, Snoopy descended to within 50,000 feet of the moon's surface.

The NASA Art Program

The age of space exploration has been documented, celebrated, exploited and even lampooned in all aspects of popular culture. The race to the moon inspired hundreds of feature films, television programs, comic books, trading cards, lunch boxes, toys, gadgets, and knickknacks, providing a boon to collectors and space enthusiasts alike.

In 1963, NASA director James Webb suggested an organized effort to commission painters, illustrators, sculptors, and other fine artists to document the dramatic events of the space program. Over the years the agency has provided small stipends to artists as celebrated as Norman Rockwell, James Wyeth, and Robert Rauschenberg.

In the 35 years since, the agency has worked with hundreds of artists and many prestigious galleries and museums to showcase the emerging genre of space-inspired art. Although many of these paintings can be found in the National Air and Space Museum, they are hardly limited to that venue. NASA-commissioned art has found its way into the Smithsonian and the National Gallery of Art as well as other prestigious collections, and should continue to inspire and educate patrons and lovers of space art for years to come.

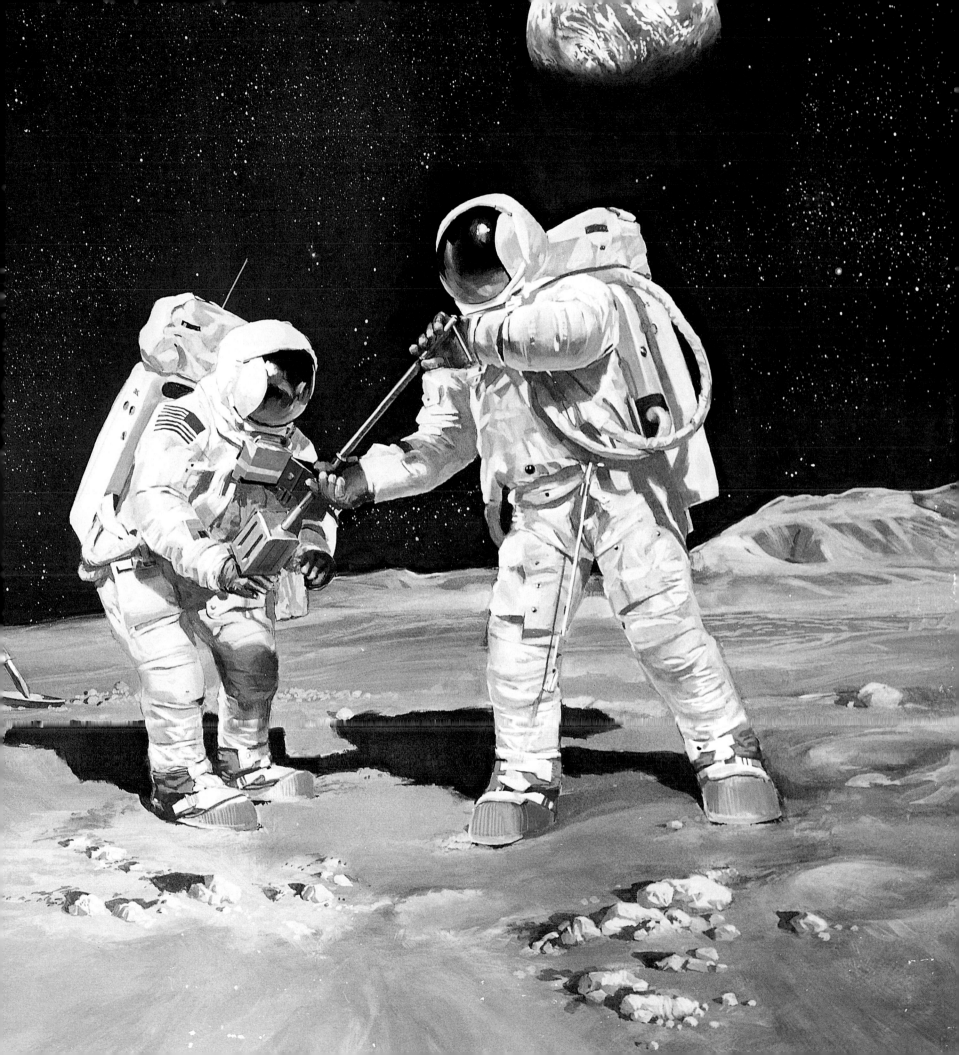

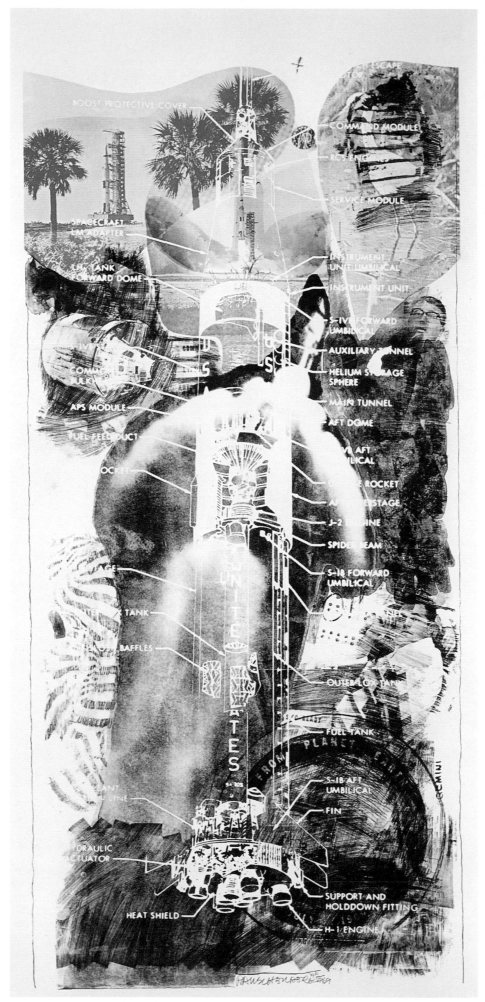

**Robert Rauschenberg's
"Sky Garden."**

A sketch by artist Robert McCall.

Three to the Moon!

NEIL ARMSTRONG

It's no exaggeration to say that Neil Armstrong was born to fly. As a teenager he worked at his town's small airport as a mechanic, and earned his pilot's license at the age of 16. He could legally fly before he could drive.

Colleagues said that his love of flight bordered on religious devotion. According to Apollo project spokesperson Paul Haney, Armstrong took a piece of the original Wright Brothers' flyer on his Gemini 8 mission, and treated it "like it was a piece of the true cross."

Born on August 5, 1930 in the town Wapakoneta, Ohio, Armstrong joined the Navy in 1949 and flew 78 combat missions during the Korean war, where he was shot down on one occasion and forced to parachute behind his own lines. After the Navy he attended Purdue and the University of Southern California, where he earned degrees in aeronautical and aerospace engineering.

In 1955 Armstrong joined a group of test pilots at Edwards Air Force Base, flying for the NACA, the National Advisory Committee on Aeronautics. Armstrong would test the X-15, the rocket plane offshoot of the X-1 jet that Chuck Yeager had used to break the sound barrier. NACA became NASA in 1959, and Armstrong flew the X-15 to an altitude of 207,000 feet in 1962. While an extraordinary altitude for a jet, it was clear to Armstrong that the future of aeronautics was in outer space. He joined the second group of NASA astronauts in September, 1962.

The Apollo 11 astronauts' wives, from left: Joan Aldrin, Patricia Collins, and Janet Armstrong.

Armstrong's experience as a fighter pilot and test pilot were put to good use during the flight of Gemini 8, when he stayed in control as his capsule began to gyrate wildly after docking with the Agena rocket. But his most acclaimed piloting would take place during the final seconds of Apollo 11's descent to the moon, when he guided the LM away from a boulder-strewn crater to a smooth landing site.

Confronted with months of preflight publicity, Armstrong maintained a sense of dignity and modesty about his role as the first man scheduled to walk on the moon. "If historians are fair," he observed. "they won't see this flight like Lindbergh's. They'll recognize that the landing is only one small part of a large program."

EDWIN E. "BUZZ" ALDRIN

Like his commander Armstrong, Edwin E. "Buzz" Aldrin was both a crack fighter pilot and a scholar before becoming an astronaut. Love of space flight was practically in his genes—his father had studied physics under Dr. Robert Goddard at Clark University in Massachusetts. Born on January 20, 1930 in the leafy suburban town of Montclair, New Jersey, Aldrin attended the U.S. Military Academy at West Point where he graduated third in his class. He flew

66 combat missions during the Korean War and was credited with destroying two enemy MIG-15s.

Aldrin earned his Ph.D. at the Massachusetts Institute of Technology, and dedicated his doctoral thesis on orbital mechanics and rendezvous to America's astronauts: "Oh, that I were one of them." In October 1963, his dream became a reality when he was chosen among the third group of NASA astronauts. Originally tapped to be on the backup crew for Gemini 10, in 1966 he was moved up to the backup crew of Gemini 9 when astronauts Elliot See and Charlie Basset were killed in a plane crash. During his Gemini 12 flight commanded by Jim Lovell, Aldrin set a new record with a five-and-a-half hour EVA.

Since leaving NASA in 1971, Aldrin has written and lectured extensively about the need for America to continue its mission in space. In his memoir he also frankly discussed the psychological toll of being an astronaut in the public spotlight.

Armstrong, Collins, and Aldrin anticipate their mission in casual attire.

MICHAEL COLLINS

While most of the world's eyes were glued to Armstrong and Aldrin as they walked on the moon, astronaut Michael Collins' mission in the command module was no less significant, since it was his job to get them all safely back to earth. He had to learn and practice more than 18 different rendezvous scenarios should the LM have to abort during its lunar descent.

While a billion or so earthlings watched Armstrong take his momentous first steps, Collins and his Columbia spacecraft were out of radio contact behind the far side of the moon. Yet Collins has written that he did not consider himself lonely or left out, but blessed with a profound experience of cosmic solitude: "If a count were taken, the score would be three billion plus two over on the other side of the moon, and one plus God knows what on this side."

Born in Rome, Italy on Halloween day in 1930, Collins attended West Point like Aldrin, and like Armstrong served as a test pilot at Edwards Air Force base during the 1950s. He joined NASA in 1963 with the third group of astronauts, and flew with astronaut John Young on the Gemini 10 mission in July 1966.

Mike Collins' post-Apollo experiences included a stint as assistant Secretary of State for Public Affairs. He went on to become the first director of the Smithsonian's renowned National Air and Space Museum, which opened in 1976.

From the Earth to the Moon: The Mission of Apollo 11

On January 6, 1969, Neil Armstrong got the nod to command Apollo 11. Given the schedule of space missions at the time, it was clear that his team would be in line to become the first to land on the moon. Astronauts Buzz Aldrin and Mike Collins were named to pilot the lunar and command modules. (Armstrong was well aware that if Apollo 9 or 10 failed, the honor would skip to the next crew.) Collins would stay in orbit while the other two men landed on the moon. The question remained, who would be first: Aldrin or Armstrong? While many have speculated that Neil Armstrong was chosen for his outgoing personality, his Midwestern background, or his status as a pilot, the real reason was far more mundane—as commander, he sat next to the exit hatch and thus was naturally the first "out the door." Indeed, it would have been awkward (and undesirable) for LEM pilot Aldrin to crawl

RIGHT: From the earliest days of the Mercury program, *Life* magazine played an essential role in selling NASA to the public. The July 25, 1969, issue.

BOTTOM: Dealing with the press was a crucial part of an astronaut's job.

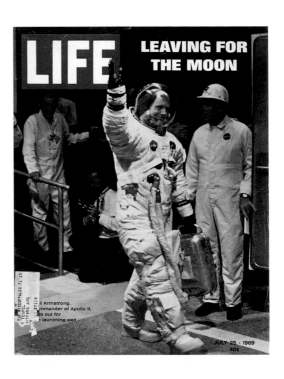

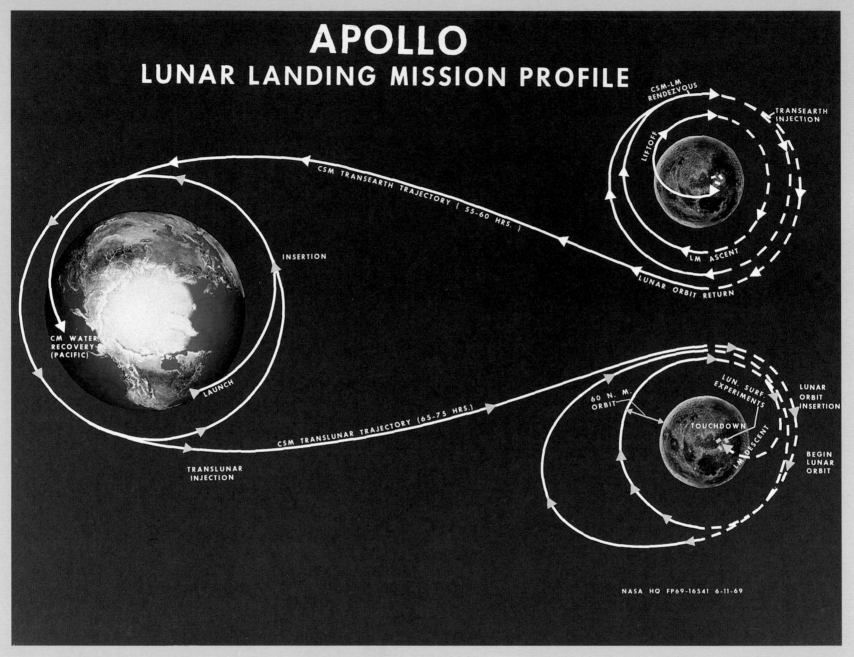

APOLLO
LUNAR LANDING MISSION PROFILE

CSM-LM RENDEZVOUS

TRANSEARTH INJECTION

LIFTOFF

CSM TRANSEARTH TRAJECTORY (55-60 HRS.)

INSERTION

LM ASCENT

LUNAR ORBIT RETURN

CM WATER RECOVERY (PACIFIC)

LAUNCH

60 N. M. ORBIT

LUN. SURF. EXPERIMENTS

LUNAR ORBIT INSERTION

TOUCHDOWN

CSM TRANSLUNAR TRAJECTORY (65-75 HRS.)

LM DESCENT

BEGIN LUNAR ORBIT

TRANSLUNAR INJECTION

NASA HQ FP69-16541 6-11-69

ABOVE: The road map for a half-million mile voyage.

OPPOSITE Armstrong practices scooping moon rocks in April, 1969.

over his commander to get out of the craft. Quite simply, Armstrong was the man.

After the first man was chosen, many wondered just what he would he say when he stepped down on the moon. Armstrong was bombarded with mail full of suggestions for quotes from Shakespeare, the Bible, and other literary worthies. The July, 1969 issue of *Esquire* magazine even ran an article in which poets, politicians, and other public figures shared their very varied opinions on the matter. But everyone agreed that Armstrong would be making history.

Everything about the Apollo 11 mission was on a gigantic, superlative scale. The Saturn 5 rocket that launched the spacecraft measured 281 feet high. With the Apollo spacecraft attached, it was 363 feet tall. The three-stage rocket was attached to the Apollo spacecraft in NASA's vast Vehicle Assembly Building (VAB), one of the largest man-made structures in the world. It covered eight acres and enclosed 129,482,000 cubic feet of space. The building itself rested on a solid foundation of 4,225 16-inch steel pilings driven 170 feet deep into the ground.

LEFT: *Esquire* queried pundits, politicians and intellectuals about what the first words uttered on the moon should be. Some worried that Armstrong would not be eloquent enough. He proved them wrong.

FAR LEFT: "The Eagle has wings."

BELOW: Mission Control.

OPPOSITE: The Saturn 5 generates 7.5 million pounds of thrust.

Once assembled, the Apollo Saturn rocket weighed 18 million pounds. It was transported three and a half miles to the launch on a six-million-pound crawler with enormous tank tracks. It moved slowly across a crawler-way as wide as an eight-lane highway.

On July 16, 1969, 3,497 journalists from 57 countries joined the half-million tourists, space buffs, and sight-seers who descended on Cape Kennedy for the Apollo 11 launch, scheduled for 9:32 a.m. Among the 7,000 dignitaries watching the liftoff were Vice President Spiro Agnew, former President Lyndon Johnson, and his wife Lady Bird Johnson. As a Senate Majority Leader, Vice President and President, Johnson had been instrumental in supporting funding for NASA and the Apollo mission. Today, the Johnson Space Center in Houston, Texas bears his name.

Compared to the jittery days of Vanguard missiles exploding on the launch pad and nervous prayers as John Glenn sat atop the Atlas-Mercury, the launch went like clockwork. *Life* correspondent Loudon Wainwright even felt a little let down by the sheer predictability of the event, writing that "precision has a way of dehumanizing adventure, even if the destination is a piece of the Moon where a man will stand."

It took less than two and a half minutes for the rocket's mighty first stage, with its 7.5 million pounds of thrust, to take the crew 220,000 feet into space before shutting down and separating. The second stage fired for another seven minutes before dropping away. By that time, the Saturn had reached a speed of 22,756.7 miles per hour. The two-and-a-half-minute burn of the third stage would bring the speed to 25,562 mph. After two hours and 44 minutes, the third stage J-2 rocket was reignited for 347.3 seconds, breaking the crew out of

BELOW: Former President Lyndon Johnson and Vice President Spiro Agnew were among the dignitaries present at liftoff.

earth's gravity and sending them on their way to the moon. Armstrong expressed his gratitude after this stage to Mission Control in Houston. "That Saturn gave us a magnificent ride," he said.

Three hours and 17 minutes into the mission, Collins fired rockets to separate the command and service module from the lunar module. The craft drifted 75 feet apart before Collins completed the delicate process of redocking. Now the astronauts and their two precious spaceships were ready for a three-day-long trip to the moon, called the "translunar coast."

Three days later, on Sunday morning July 20, 1969, the lunar module Eagle separated from Columbia. "You cats take it easy on the lunar surface," advised Michael Collins. Commander Neil Armstrong radioed Houston, "The Eagle has wings!" All the while, Collins inspected the departing LM to check that all its landing gear was intact and attached for its descent of more than 60 miles to the lunar surface. "I think you've got a fine-looking flying machine down there, despite the fact that you're flying upside down," joked Collins.

An hour after undocking, Armstrong fired the first descent engine that would take the Eagle down to a mere 50,000 feet over the moon. Five minutes into the burn and only 6,000 feet in altitude, alarm bells began ringing, indicating computer malfunctions. Another bell rang at 3,000 feet. But still the Eagle descended. Despite the distraction of the alarm bells, at 1,000 feet Armstrong could see their planned landing site, a large crater containing boulders the size of a compact car.

With only 60 seconds of fuel left and with the LM's computer and radar providing mixed and confusing signals,

BELOW: Thousands of ordinary citizens waited to catch a glimpse of the historic launch.

OPPOSITE: Aldrin descends from the LM.

Armstrong and Aldrin began to see a cloud of lunar dust being kicked up by the rocket's thrust. Armstrong was so concerned by the dust and lack of visibility that he failed to notice when the Eagle finally landed with only a few seconds' supply of fuel remaining. Although hardly intended for posterity, the first words said on the moon's surface were from Buzz Aldrin: "Okay, engine stop." Armstrong followed with words that still ring with history, "Houston, The Eagle has landed."

The reply from earth was about as emotional as Mission Control would allow. "Roger, Tranquility. We copy you on the ground. You've got a bunch of guys about to turn blue. We're breathing again."

OPPOSITE: Buzz Aldrin contemplates Old Glory on the surface of the moon.

BOTTOM Mission Control: "We're breathing again."

Following the drama of their descent, the two astronauts took a six-and-a-half-hour "breather" in the cramped lunar module. Buzz Aldrin used some of the time to commemorate the historic moment by performing a personal communion service using a chalice, wafer, and wine he had brought in a small plastic bag. It took longer than scheduled to suit up for their moon walk, so it was already 10:28 p.m. Eastern Time when Armstrong activated the TV camera that would broadcast his first steps to an estimated one billion viewers on earth. As he touched down, he said the words he knew would go into the history books: "That's one small step for man. One giant leap for mankind."

But he said them incorrectly. He had intended to say "that's one small step for *a* man," though that's not quite how it came out.

After a few minutes on the moon, Armstrong could not help but be moved by its desolate landscape. "It has a stark beauty all its own," he observed, sounding more like a poet than a pilot. After fourteen minutes of ceding the spotlight to his co-pilot, Aldrin asked, with anticipation, "Are you ready for me to come out?" And when he did, the poetry continued.

Aldrin: "Beautiful, beautiful."

Armstrong: "Isn't that something. Magnificent sight down there."

Aldrin: Magnificent desolation."

Back on earth, pundits were equally speechless. At one moment, the usually voluble news anchor Walter Cronkite had little to say. "Man has landed and Man has taken his first steps. What is there to add to that?"

LEFT: Aldrin at the foot of the LM. A good indication of the size of the lunar module.

BOTTOM: Aldrin sets up an experiment measuring solar wind.

NEW GLORY FOR OLD GLORY

Together, Armstrong and Aldrin tested the effects of walking, bounding, and jumping in the weak lunar gravity. They unwrapped the protective cover from a stainless steel plaque attached to the Eagle's front landing gear, and Armstrong read the commemorative words for his audience on earth: "Here men from the planet earth first set foot on the moon July 1969, A.D. We came in peace for all mankind."

Armstrong then removed a TV camera from beneath the LM to provide a panoramic view of the lunar horizon. Aldrin set up an instrument to measure solar winds. The two men then planted an American flag which had a built-in wire stiffener to give the effect that it was blowing in the wind, which was of course impossible since the moon has no atmosphere to provide a breeze. Both men had difficulty penetrating the lunar soil with the flagpole to a depth of more than four or five inches. As a result, the flag was raised at a slight angle when Aldrin gave it a silent salute.

OPPOSITE: Aldrin sets up the Passive Seismic Experiment Package.

LEFT: This July, 1969, New York *Daily News* cartoon by Warren King reflects American pride in the Apollo 11 mission.

BOTTOM: The astronauts left a good deal of equipment behind on the lunar surface.

ABOVE: Armstrong and Aldrin take a call from the Oval Office.

OPPOSITE: Columbia splashed down within 13 miles of the recovery ship.

At this moment, the astronauts were asked to stand together in front of the cameras as Houston put through a phone call from President Richard Nixon:

"Hello, Neil and Buzz. I am talking to you by telephone from the Oval Office of the White House. And this certainly has to be the most historic telephone call ever made. I just can't tell you how proud we are of what you have done. For every American this has to be the proudest days of our lives . . . because of what you've done, the heavens have [become] a part of man's world and as you talk to us from the Sea of Tranquility, it inspires us to double our efforts to bring peace and tranquility to earth. For one priceless moment in the history of man, all the peoples on this earth are truly one. One in their pride in

what you have done, and one in our prayers that you will return safely to earth."

With this ceremony over, the two men returned to their busy lunar schedule. As Aldrin photographed Tranquility Base, Armstrong collected random "undocumented" rock samples within a 98-foot radius of the LM. Armstrong also positioned a laser reflector which beamed signals back to tracking stations on earth. Aldrin set up a permanent, extremely sensitive seismic measuring device to track meteorite impacts, "moonquakes," and to determine the interior composition of the moon. To withstand the brutal cold of lunar nights (where temperatures can fall as low as -279 degrees Fahrenheit), the device contained a heating system powered by 1.2 ounces of plutonium.

Aldrin and Armstrong then took more than 20 trips around the site's perimeter to collect "documented" moon rocks and soil samples that were placed in specially designated tubes and boxes. Together they collected 48.5 pounds of material.

Just before reentering the LM, Aldrin took time to drop a small sack on the moon. It contained mission patches memorializing the Apollo 1 astronauts Gus Grissom, Ed White, and Roger Chaffee, and two medals to commemorate the late Soviet cosmonauts Yuri Gagarin and Vladimir Komarov. Armstrong then followed Aldrin into the LM, but not before wiping his dusty moon boots off on the forward foot pad. Once both men were back in the Eagle, they discarded some unnecessary items including backpacks, a camera, space suit connector covers, a lunar tether, urine bags, and two pairs of moon boots. Because their boots were considered the most likely to be contaminated by any moon "germs," they were zipped into a special bag once the astronauts reentered the LM. The highly sensitive seismic measuring system that Aldrin had just set up began to register its detection of the cast-off materials. Man had now officially left litter behind on the surface of the moon.

After the garbage detail, the astronauts repressurized the LM and settled in for a mandatory rest period. Sleep proved elusive for the two moon walkers. The cramped, chilly Eagle was filled with the noise of fans and equipment, and reflected light from the moon's surface and the earth itself shone through the window shades.

Finally, 21 hours and 36 minutes after landing on the moon, the ascent stage of the Eagle was set to take off. While not as dramatic as the first landing, it was just as crucial. If the LM's rockets failed, the astronauts would be left to die on the moon. If Aldrin was nervous, however, he hid it well with his sense of humor:

Houston: "Our guidance recommendation is PGNCS and you are cleared for takeoff."

Aldrin: "Roger, understand. We're number one on the runway."

The ascent rocket worked perfectly, lifting the Eagle into orbit for a rendezvous with Columbia. After a brief

RIGHT: The astronauts would not leave quarantine until August 10, 1969.

eight-second scare while the two spacecrafts gyrated out of alignment, the three men went about the task of transferring the precious cargo of moon rocks from the Eagle to Columbia. Once emptied, the Eagle was separated from Columbia and abandoned in moon orbit. Then Columbia fired its rockets to escape the moon's gravity and begin the three-day coast towards earth.

On July 24, 1969, Columbia separated from the service module and entered the earth's atmosphere traveling at 24,705 miles per hour at an altitude of 400,000 feet. At 24,000 feet, the first of three sets of parachutes were deployed to allow Columbia to splash down at the relatively gentle speed of 31 feet per second. After a voyage of a half-million miles, the spacecraft landed within 13 miles of the recovery ship, the aircraft carrier Hornet carrying President Richard Nixon who would personally address the returning heroes.

But he could not shake their hands!

Once retrieved from the bobbing waters of the Pacific, the three men were immediately dressed in gray biological isolation outfits and housed in a quarantine trailer. They may have returned safely to earth, but they looked like creatures from another world as television cameras captured them taking their first awkward steps in nonlunar gravity.

Three days later, they were returned to Houston in a mobile quarantine facility. There they were joined by experimental mice who were to be exposed to the moon "germs," if any, that the astronauts had brought back. It wasn't until August 10, more than three weeks after blastoff, that the three were finally able to return to their wives and families and to receive the thanks of a grateful nation.

Armstrong, Aldrin, and Collins were received as conquering heroes on a junket of parades and parties. At one such gathering an astronaut raised a glass and gave a toast. "Here's to the Apollo program. It's all over." He was more than half right. Five more Apollo missions would follow, and four would reach and explore the moon. But his bittersweet remark spoke volumes about the public and political attitudes that would later ensue towards the space program.

BELOW: President Richard Nixon applauds the returning trio.

Eight Miles High: Pop Culture Spaces Out

When Apollo 11 reached the moon, Mission Control in Houston flashed the words of John F. Kennedy, proclaiming the goal of putting a man on the moon before the decade was out. These words were followed by the phrase "Task Accomplished, July 1969." While the engineers on the ground had a sense of a job well done, the public's attitude was more accurately summed up by a hit Peggy Lee pop song of the period. "Is That All There Is?"

NASA was hardly immune to the enormous social changes that engulfed America in the 1960s. While Neil Armstrong was

Space proved inspiration for fads in music, fashion, food and toys. surely sincere when he read the words "we come in peace for all mankind" on the moon, for many on earth, particularly the young, NASA was part of the same wartime "establishment"

that was responsible for the era's tragic debacle in Vietnam.

Changing attitudes towards the space program were reflected in popular culture during the tumultuous decade. During the heady, early days of the Mercury program, NASA's go-go spirit was positively reflected in pop music. The guitar band the Ventures had a 1962 hit with "Telstar," the name of a communications satellite. They named one of their albums "Blast Off." Another twang-and-surf band even called themselves The Astronauts.

But by mid-decade, the times and tunes they were a-changin'. In September 1965, Barry McGuire had a number-one hit with the Dylan-esque

protest record "Eve of Destruction." This angry, doom-laden ditty challenged, "You can leave the world for four days in space/But when you come back it's the same old place." The song echoed a growing sentiment that missions to the stars were pointless folly with the planet earth in such dire shape.

By the late '60s, the image of the astronaut

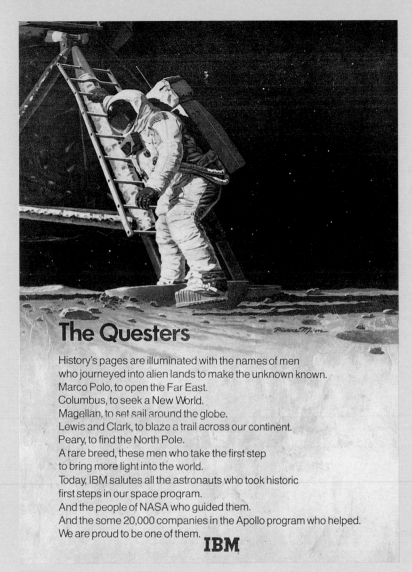

The Questers

History's pages are illuminated with the names of men
who journeyed into alien lands to make the unknown known.
Marco Polo, to open the Far East.
Columbus, to seek a New World.
Magellan, to set sail around the globe.
Lewis and Clark, to blaze a trail across our continent.
Peary, to find the North Pole.
A rare breed, these men who take the first step
to bring more light into the world.
Today, IBM salutes all the astronauts who took historic
first steps in our space program.
And the people of NASA who guided them.
And the some 20,000 companies in the Apollo program who helped.
We are proud to be one of them. **IBM**

in popular music was no longer heroic. The Rolling Stones' "2000 Lightyears From Home" and David Bowie's first hit "Space Oddity" both depicted space travel as a metaphor for loneliness and alienation. The psychedelic music craze of the mid to late '60s used images of flight and space as obvious drug allusions. In the Byrds' popular "Eight Miles High" and "Mr. Space Man," getting high was clearly not a reference to altitude.

The week Neil Armstrong walked on the moon, the number-one single on the pop charts was "In the Year 2525" by Zager and Evans. The driving, repetitive song portrayed a bleak future in which all human passions and appetites have been banished by technology. Less than a month after Apollo 11, more than 400,000 music fans converged on Bethel, New York for the Woodstock music festival. While some saw Woodstock as a mere concert and bacchanalian blast, many in the media construed the event as a portent of a new generation, a "Woodstock Nation" committed to moving America away from its obsession with materialism, militarism, and technology. In the words of Joni Mitchell, who wrote a song about the event, it was time to "get back to the garden."

Corporate giants like IBM and McDonnell Douglas were proud to point out their links to Apollo's success.

From Freedom 7 to Apollo 11
...our eight years to the Moon.
Beginning with the pioneering "Freedom 7" mission of 1961, we have been a part of all 21 U.S. manned space flights. We built NASA's Mercury and Gemini spacecraft, and the S-IVB stage which orbited the Apollo spacecraft and thrust the historic Apollo 11 to the Moon. Today, as a company dedicated to space science and exploration, we're working toward still greater achievements, in Earth-orbit, and beyond.

MCDONNELL DOUGLAS

Promise and Anti-Climax: The End of Project Apollo

The flight of Apollo 12's Intrepid and Yankee Clipper in November 1969 was no less dramatic than its predecessor. Astronauts Pete Conrad and Alan Bean returned with almost twice as many moon rocks as Apollo 11 and spent 31.5 hours on the moon's Ocean of Storms. They even returned with parts of the unmanned Surveyor 3 spacecraft which had landed more than two years earlier, yet by now public interest in the space program had dramatically declined.

Sadly, a public that had learned to take moon missions in stride received a shock with the mission of Apollo 13. The craft was nearly halfway to the moon when an oxygen tank in the command module Odyssey exploded. Short on both fuel and oxygen, the three crewmembers were forced to ride in the lunar module Aquarius. The world held its breath as the wounded craft entered lunar orbit and then limped its way back 250,000 miles to earth. When the crew successfully splashed down, NASA official Robert R. Gilruth commented that Apollo 13 was a scary reminder that "flying to the moon is not just a bus ride."

New York City saluted Armstrong, Aldrin and Collins with a ticker tape parade. Their limousine hardly seems space age.

After the nail-biting drama of Apollo 13, the mission of Apollo 14 between January 31 and February 9, 1971, would be remembered for one of the program's lighter moments, when astronaut Alan Shepard took time to play golf on the moon's surface. The last three Apollo missions took advantage of a larger lunar module that allowed longer lunar stays. They also employed the foldable, four-wheel drive lunar roving vehicle (LRV) or "Rover" for short. The Rover had a top speed of seven miles per hour and a battery that enabled it to travel 55 miles. The astronauts were forbidden to drive more than six miles from their ship, allowing them to walk "home" should the Rover break down. The moon "buggy" allowed Apollo 15 astronauts David Scott and James Irwin to explore for more than 19 hours and retrieve 169 pounds of rocks. Apollo 16 would spend even longer on the moon, and Apollo 17 would accomplish a record four EVAs totaling 22 hours.

When Apollo 17 ascended from the lunar surface on December 7, 1972, it left behind a plaque with a map of the earth and the moon with the sites of the Apollo landing sites clearly marked. An inscription read "Here man completed his first exploration of the Moon, December 1972."

Begun with much fanfare, the Apollo program ended with more of a whimper than a bang. With its budgets slashed, NASA canceled the missions of Apollo 18, 19, and 20. Skylab, the first post-Apollo program, also saw its mission curtailed. This low-altitude space station housed three missions in 1973 and 1974 before being abandoned. The last Skylab mission spent 84 days in space, a record that would last for years. By a strange quirk of fate, the station crashed back into the earth's atmosphere on July 11, 1979, a rather pathetic reminder of the state of America's space efforts only nine days before the 10th anniversary of Apollo 11.

Down to Earth: Spinoffs of the Space Program

Thirty years after Neil Armstrong's historic walk on the moon, there are an estimated 30,000 down-to-earth by-products and technological spinoffs that resulted from Project Apollo and the space program. From the garage to the hospital, from the kitchen to the golf links, NASA-derived technology has revolutionized modern life. To mention but a few:

- A pen that works in zero gravity and upside down.
- The coating for scratch-resistant eyeglass lenses was originally developed to protect space gear.
- The breathing apparatus used on Apollo space walks now helps firefighters stay alive.
- Shock absorbing-materials found in today's sneakers and running shoes uses technology developed for the Apollo space suits.
- Black & Decker first created the cordless drill to dig soil samples on the lunar surface.
- The joy stick developed for the Lunar Rover now allows quadriplegics to drive cars.
- A self-righting raft, developed to keep the Apollo capsule afloat after splashdown, has saved more than 500 lives in 10 years.
- Liquid circulating long johns which kept astronauts from stifling in their space suits have been adapted for professionals, including race car drives, nuclear reactor technicians, and shipyard workers. Four-hundred of these "cool suits" were used during the Gulf War.
- Smoke detector technology found in almost every home was originally made for NASA.
- Pacemakers use telemetry devices first used to monitor astronauts.
- NASA-developed aluminized materials used to insulate satellites and manned spacecraft have been adapted into lightweight reflective "space blankets" used for survival in harsh conditions, as well as for the insulation of water heaters and houses.
- Lightweight composite materials made of a mixture of resins and fibers have found their way from aerospace applications to lightweight bike helmets, tennis rackets, and golf clubs.
- The quartz timing crystals that revolutionized the watch industry in the 1970s were first developed for NASA.
- The bar code used to identify virtually every consumer product was first employed by NASA to maintain an inventory of countless spacecraft parts.

So what exactly is the legacy of Project Apollo? In addition to its technological innovations, Apollo proved to America and the world that great and difficult tasks can be accomplished when enough time, resources, and ingenuity are employed. Like the Panama Canal and the Manhattan Project, Apollo was an inspiring example of American vision and determination and a remarkable triumph of management and logistics. John F. Kennedy's words, uttered at the birth of the program, best sum up the spirit of Apollo:

"We choose to go to the moon and do the other things, not because they are easy but because they are hard. Because that goal will serve to organize and measure the best of our energy and skill. Because that challenge is one that we are willing to accept, one that we are unwilling to postpone, and one we intend to win."

At the same time, Apollo's greatest and most profound gift to mankind was probably entirely unintentional. Only by traveling to the moon could people on earth finally see the planet in its beauty and entirety. The photographic image of the earth as a "big blue marble" floating in space was a powerful inspiration for the first "Earth Day" celebration on April 22, 1970, and for the three decades of environmental awareness that have followed. As Apollo 8 astronaut William Anders reflected, "We came all this way to explore the moon, and the most important thing is that we discovered the earth."

ABOVE: Space blankets and Fisher antigravity pens were among the thousands of products inspired by the space program.

NEXT PAGE: It took a trip to the moon to make some appreciate the beauty and fragility of the planet earth.

Bibliography

Arnold, H.J.P., ed. *Man in Space*. New York: Smithmark, 1993.

Bonestell, Chesley and Ley, Willy. *The Conquest of Space*. New York: Viking Press, 1949.

CBS News. 10:56:20 p.m. 7/20/69. New York: Columbia Broadcasting System, 1970.

Chaikin, Andrew. *A Man on the Moon*. New York: Penguin, 1993.

Clarke, Arthur C., *The Exploration of Space*. New York: Harper and Brothers, 1951.

Dickey, Beth. *Nasa Technology Spin-offs*. New York: ABC News.com, 1998.

Fritsche, Alan, G., ed., *The Flight of Apollo Eleven*. Memphis: STS Mission Profiles, 1994.

Garber, Stephen J. and Launius, Roger, D. *A Brief History of the National Aeronautics and Space Administration*. Washington: NASA History Office, 1998.

Halberstam, David. *The Fifties*. New York: Villard, 1993.

Launius, Roger, D. and Ulrich, Bertram. *NASA and the Exploration of Space*. New York: Stewart, Tabori & Chang, 1998.

Ley, Willy. *Rockets and Space Travel*. New York: Viking Press, 1947.

McAleer, Neil, *The Omni Space Almanac*. New York: World Almanac, 1987.

McCurdy, Howard E. *Space and the American Imagination*. Washington: Smithsonian Institution Press, 1997.

Murray Charles, and Cos, Catherine Bly. *Apollo: The Race to the Moon*. New York: Simon & Schuster, 1989.

Neal, Valerie, and Lewis, Cathleen S. *Spaceflight: A Smithsonian Guide*. New York: Macmillan, 1995.

Shelton, William, R. *Man's Conquest of Space*. Washington: National Geographic Society, 1968.

Weldon, Michael. *The Psychotronic Encyclopedia of Film*. New York: Ballantine, 1983

Wolfe, Tom. *The Right Stuff*. New York: Farrar Straus & Giroux, 1979.

Photograph and Illustration Credits

Metric Conversions

Temperature			Length		
When you know	Multiply by	To find	When you know	Multiply by	To find
°F / Fahrenheit temp.	5/9 (-32)	Celsius temp. /C°	in./inches	2.54	centimeters /CM
°C / Celsius temp.	9/5 (+32)	Fahrenheit temp. /F°	ft./ feet.	30	centimeters /CM

Index